LIFE

Birds

THE LARGEST OF THE North American herons, the great blue heron stands above all others.

Contents

PAINTED BUNTINGS ARE known for their vivid fusion of blue, green, yellow, and red feathers. The deep-blue indigo bunting (near left) is one of the most abundant songbirds in the eastern United States.

The Colors, the Songs, the Majesty of Flight

Why Humans Love Birds

BY MELISSA MAYNTZ

THE FIRST BIRD I FELL IN LOVE WITH—MY "SPARK BIRD"—WAS SOARING IN THE NORTHEASTERN Florida sky one May many years ago, its pointed wings spread gracefully and its deeply forked tail gently twisting to guide its curving flight. The black-and-white coloration was distinctive, and I rushed to a local used bookstore for what would become the first of many field guides on my shelves. A few flips of the pages, and I knew I'd seen a swallow-tailed kite, an elegant raptor that is a summer visitor to Florida and the southeastern United States.

Tens of millions of birders have had similar encounters with their own spark bird. In the United States alone, more than 45 million people are bird watchers. Roughly $4 billion is spent annually

AN ESTIMATED 50 BILLION birds—more than 10,000 distinct species—populate the planet. Such diversity inspires tens of millions of birders in North America, where the bald eagle soars.

on birdseed and foods such as suet, nuts, and nectar, while another $2 billion is spent on binoculars, spotting scopes, and other equipment. Birds are the focus of conservation programs and citizen science projects such as the Great Backyard Bird Count; art projects like the Audubon Mural Project in New York City, which highlights 314 bird species; and movies like *Happy Feet* (about penguins) and *The Big Year* (about a birding competition).

That so many people love birds may be partly because there's a bird for everyone. The more than 10,000 known bird species come in an extraordinary variety, and they can be found—almost literally—everywhere.

Birds thrive in all habitats, from fierce roadrunners in rocky deserts to colorful toucans in tropical jungles. You don't need to live next to a wildlife refuge or nature preserve to enjoy a multitude of bird species—even the busiest cities are home to swallows and sparrows, hawks nesting on skyscrapers, ducks in park ponds, and hummingbirds in flower beds. Taking a trip to the beach? Watch for sandpipers running from the waves, pelicans floating on the water, and gulls flocking on the dunes. In rural areas, there might be quail, magpies, and wild turkeys at the edges of farm fields, while suburban yards can be flush with thrushes, warblers, and buntings. Wherever we are, birds provide us with an active, living connection to nature.

Birds' often vibrant colors can distinguish a species in a beautiful way. The brilliant red of the northern cardinal stands out against winter snows, while the bright hue of the blue jay is a bold splash of color among the leaves. Birds come in every color, and some—like the painted bunting, with his blue head, lime-green back, and rich red chest—are a rainbow all by themselves.

The varied hues have a purpose. Brighter colors can help birds attract stronger mates—as a general but not ironclad rule, the more colorful of the species are the males, with female birds often more muted, more demure, in their coloring. In other cases, a mottled

pattern provides camouflage to protect nesting birds, and some birds, such as the northern pygmy owl, even have false "eyespots" on the back of their head to fool potential predators.

THE SIGHT OF BIRDS IN FLIGHT SUGGESTS a sense of freedom—from the awesome dive of a peregrine falcon to the swooping curves of a barn swallow, or even the quick flitting of a house wren. The long-distance migrations of birds, flying hundreds or even thousands of miles between their summer and winter habitats, highlight their endurance and perseverance, as well as a kind of navigational intelligence. Birds rely on landmarks and stars to guide their journeys.

There's a variety in how they fly as well. Consider the hours-long flights of albatrosses out at sea as they soar on air currents; the frantic, adrenaline-inducing flights of pheasants scattering from predators; or the flittering flights of foraging warblers navigating the high trees without hitting a single branch.

And of course, there's their songs. Chirps, whistles, coos, and warbles are as familiar as the somewhat less melodious screeches, squawks, hoots, and quacks. Some birds, such as mockingbirds, thrashers, and catbirds, are outstanding mimics and imitate not only other birds but also other animals—as well as car alarms and ring tones.

Birds sing to attract mates and to defend their territory, with more complex songs indicating better health and greater experience to lure the very best mates or defend larger territories. Other songs and calls communicate information about food or predators, and while in flight, flocks of birds often call to one another to maintain proper spacing with their airborne neighbors.

The reasons why humans appreciate birds are almost as diverse as birds themselves. The bill of a roseate spoonbill, the hovering of a hummingbird, the gleam of an eagle's eye, the trill of a nightingale in the gloaming: Maybe one of those birds is *your* spark bird, long since catalogued or quite literally just up around the bend. ●

Bold and Beautiful

Some birds are as brightly colorful as anything in the natural world. And each has a story of its own.

HUMMINGBIRD

Found only in the Americas, hummingbirds are recognizable by their brilliant colors, compact size, and seemingly vibrating wings. A hummingbird's dazzling throat color is caused not by feather pigmentation but rather by iridescence in the arrangement of the feathers. The vividness and color tone of the throat can be influenced by a number of factors, including light level, moisture, wear and tear, and angle of viewing.

SCARLET MACAW

With a creamy white face and bright red plumage, the scarlet macaw is one of the most recognizable parrots in the world. Often found in rain forests, these birds spend their time in tall trees near rivers, usually congregating in large, noisy groups. At nearly three feet long, with a wingspan of almost four feet, scarlet macaws are one of the largest of the parrots. Almost half of their length is made up of their deep blue and sunny yellow tail.

FLAMINGO

With their distinct stance and craned necks, pink flamingos are one of the most recognizable species of waterfowl in the world. The red, yellow, or orange pigments that define flamingos' plumes come from what they eat. Many plants produce carotenoids that give carrots their orange color or turn tomatoes red. These compounds are also in certain algae that brine shrimp eat. A wild flamingo's diet includes these shrimp and algae, which causes the carotenoids to metabolize and create those fiery-colored feathers.

PEACOCK

The unmistakable male Indian peafowl, also known as a peacock, spreads its intricate, ornamental tail feathers when courting peahens. The female birds are much less flashy than their iridescent male counterparts, whose mating ritual is called "train rattling." After every mating season, a peacock sheds its feathers as part of a molting process, regrowing them longer and fuller in time for the next mating season.

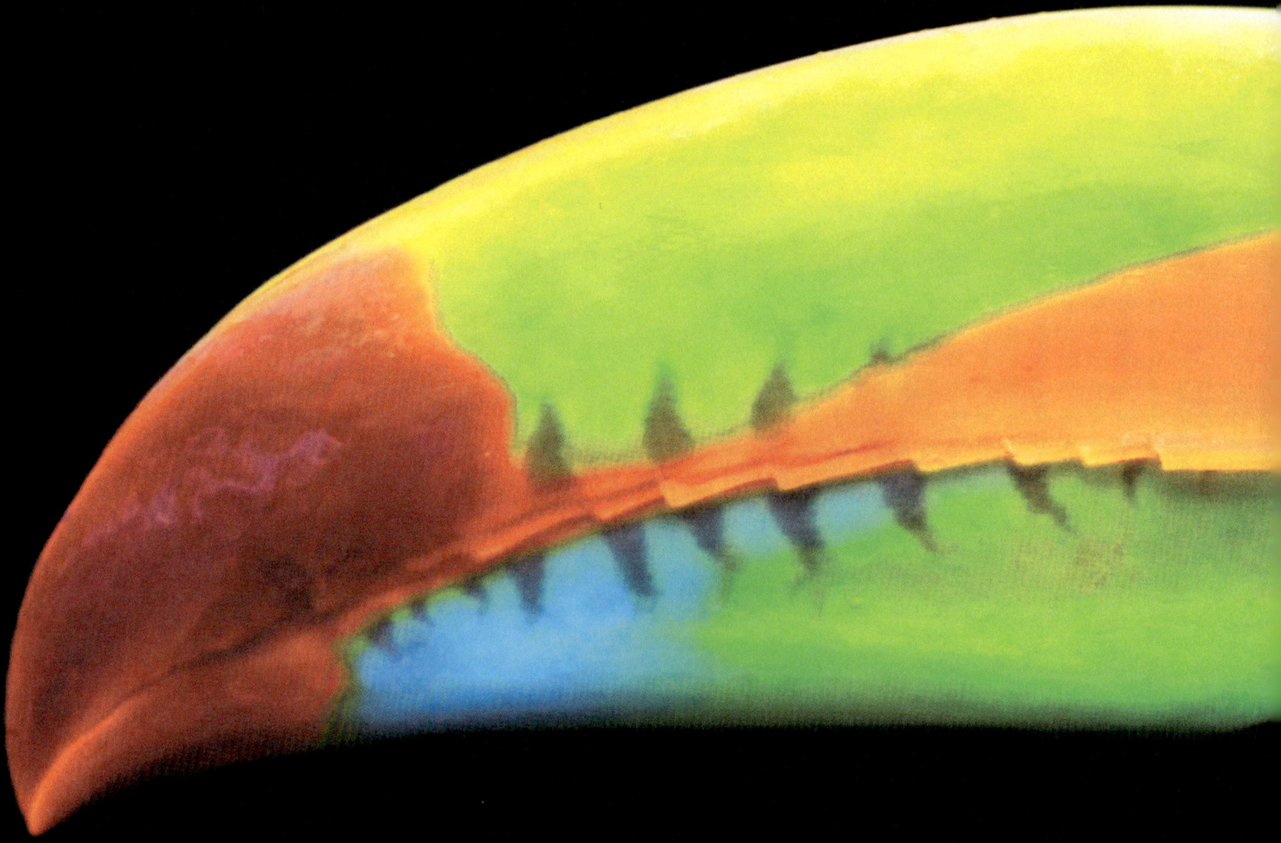

TOUCAN

Despite its size, the keel-billed toucan's vibrant, banana-shaped bill is relatively lightweight. Made of the protein keratin, the hollow beak is supported by thin bone rods. However, the toucan's broad wings are heavy for its body size, which makes flying through forests a chore. Instead, it usually hops from branch to branch through the canopy of the rain forest. These birds also have a frog-like call that can be heard up to half a mile away.

MANDARIN DUCK

The beautiful, unmistakable mandarin duck hails from China, Japan, Korea, and parts of Russia. In spring and early summer, males, such as the one at right, have elaborate, colorful plumage. Females are a little less eye-catching, with gray feathers and a muted bill. After the mating season, the males' feathers molt to brown and gray as well. By fall, they usually regain their colors.

RAINBOW LORIKEET

The vibrant rainbow lorikeet can be found in the wild in Indonesia, Polynesia, Australia, and Papua New Guinea. Most live in the rain forest canopy or at the edge of wooded areas near open savannahs. There are about 53 species of lorikeets, and they are known for their sweet nature and talkative personality.

NEW CALEDONIAN CROWS live on clusters of remote islands in the South Pacific. They are well known to scientists for their ability to make and use tools. These crows prefer hooked-stick tools, seen here, to pry pieces of meat from logs.

How Birds Think

Judging animal intelligence can be a challenge, but birds have ways of showing their smarts.

BY DAVID BJERKLIE

"What was going on in that raven's mind!?" It's a question that still intrigues Laura Erickson four decades after her encounter with the bird. Erickson, who has written more than a dozen books on birds and won a slew of awards from ornithology societies and birding associations, was on a long walk along a quiet country road. On the same path the day before, her wristwatch, a treasured gift from her husband, had slipped off, and she feared it had been lost forever. But now, a few miles into her walk, she heard a raven squawking. "A few minutes later, the raven flew over with something hanging out of its beak," Erickson says. "I pulled up my binoculars, and there was my wristwatch! The raven flew in closer and closer, right over my head. And voilà! It dropped my wristwatch at my feet." As Erickson, who hosts the long-running radio program *For the Birds*, later recounted, "I'm still gobsmacked. I've spent the intervening years studying everything I could about avian intelligence, but the more I learn, the more I realize that I haven't even scratched the surface."

Bird brains are being explored like never before, and through a fresh lens. As Emory University biologist and primatologist Frans de Waal, PhD, writes in the foreword of *Bird Brain: An Exploration of Avian Intelligence*, "Despite a long tradition of research, scientists have always carefully avoided the term *cognition* when it came to birds." Why? Because birds lack what is considered the neurological crown jewel of intelligence: the prefrontal cortex.

"Having feathers, it was concluded, simply couldn't go together with advanced learning, let alone thinking," de Waal explains. As a consequence, "birds were lumped in with fish and insects as organisms driven by instinct."

That began to change, however, when anatomists realized that birds had their own functional equivalent to the prefrontal cortex: the nidopallium, where neurons are tightly packed and wiring is minimized for maximum efficiency (in terms of size, the brain-to-body ratio of smart birds is comparable to the ratio in smart primates). Researchers now focus on how birds solve an array of challenges, environmental as well as social, and the result, according to de Waal, "has revolutionized our appreciation of what birds are capable of."

Defining intelligence can still be tricky, explains Nathan Emery, PhD, a senior lecturer in biological and behavioral sciences at Queen Mary University of London and the author of *Bird Brain.* Still, the question these days is not whether birds are intelligent but rather how and why they are.

Among the factors driving the evolution of bird intelligence are the demands of living in a social group. For birds, Emery writes in *Bird Brain,* "it is the quality of relationships, not the quantity, that calls for additional brainpower." The mental challenge is not coping with the dynamics of large flocks. "The really demanding task—at least from a psychological and cognitive point of view—is forming close alliances, especially forging bonds with a mate and providing long-term parental care to young," Emery says.

Ecological circumstances can also foster adaptation. In his book *Darwin Comes to Town: How the Urban Jungle Drives Evolution,* Dutch ecologist Menno Schilthuizen, PhD, cites examples of animals meeting the challenges—and opportunities—of living in close quarters with humans. Schilthuizen details how, in 1995, , the zoologist Yoshiaki Nihei at Tohoku University in Sendai reported that crows were using moving cars as nutcrackers. "The crows would wait near a traffic light, wait for it to turn red, and

THROUGHOUT THE PAST decade, scientists have been studying how crows use tools (below). Researchers have found that the birds aren't born with an innate ability to use tools but can learn and then pass that knowledge down to their young. Right: Over 20 years of training, animal psychologist Irene Pepperberg, PhD, taught an African gray parrot named Alex how to identify more than 50 objects.

then step in front of the cars, place their nuts, and hop back to the curb to wait for the light to change." When the traffic had passed, the crows retrieved their booty. In England, small songbirds called tits learned to pilfer the layer of cream that collected on the top of glass bottles of milk. The behavior began in the late 19th century, "when milkmen were still in the habit of leaving open bottles of milk on people's doorstep in the morning," Schilthuizen explains. The tits would swoop in, land on the bottles, and drink the cream. In some areas, milk suppliers eventually capped the bottles, first with cardboard, then aluminum. But the resourceful tits learned to overcome both obstacles and have their cream nonetheless.

RESEARCHERS NOW FOCUS ON HOW BIRDS SOLVE CHALLENGES, WHICH HAS IMPACTED OUR APPRECIATION OF WHAT BIRDS ARE CAPABLE OF.

THERE'S A TREASURE TROVE OF examples of such avian ingenuity. In the Galapagos Islands, the woodpecker finch uses stick tools to probe for insects under the bark of trees. Not only do the finches break off twigs, leaf stems, or cactus spines to use as tools, they keep their best tools for reuse. "Another classic example of avian tool use is heron bait-fishing," Emery notes. The heron "will grasp small insects in the tip of its bill and hold them over the surface of the water. Fish attracted to the insects rise to the surface, becoming targets. Burrowing owls do something similar, placing dung at the entrance to their burrow to attract insects such as beetles, which are then eaten."

The star of avian tool use, however, is the New Caledonian crow, named after the islands in the South Pacific where it is found. The crows make two types of tools. The first is made by excising narrow strips from pandanus leaves and is used to probe for prey on the ground. The second is a hooked stick tool that the crows fashion by selecting an appropriately shaped branch and then detaching smaller twigs and leaves, in effect sculpting the tool. This one is used to poke insect larvae out of tree bark.

New Caledonian crows have also strutted their stuff in laboratory experiments devised to test their understanding of how tools work. The most impressive example of such understanding, notes Emery, was demonstrated by a female crow called Betty, who bent a straight piece of

ACORN WOODPECKERS LIKE the one seen here drill holes in dead trees to hoard acorns. In the fall the birds harvest acorns to store in these holes, which provide food during the barren winter. Opposite, top: A male satin bowerbird decorates its shelter with bright-blue utensils to attract a mate. Opposite, bottom: Blue tits in England compete for the top of the cream on a milk bottle.

metal wire into a hook in order to pull up a tiny bucket containing food placed at the bottom of a test tube. Northern blue jays have also been tested in the lab, where they have successfully used feathers, thistles, straw grass, paper clips, and plastic bag ties to reach otherwise inaccessible food pellets. Researchers have even re-created "The Crow and the Pitcher" from *Aesop's Fables*. In the lab version, says Emery, a rook, a close cousin of the crow, is presented with a tube half full with water with a tasty worm floating on the surface. But how to reach the worm? The rook starts placing stones into the tube, which raises the water level . . . until the worm moves into beak reach.

BURROWING OWLS USE DUNG AT THE ENTRANCE OF THEIR BURROW TO ATTRACT INSECTS LIKE BEETLES, WHICH ARE THEN EATEN.

Clever, indeed. In 2004, the phrase *feathered apes* was introduced to make the point "that some members of the corvid [crow] family demonstrate cognitive skills that are on a par with those of the great apes," says Emery. There are corvid champions who push the primate association even farther. In their book *Gifts of the Crow*, University of Washington professor John Marzluff, PhD, and his coauthor, artist Tony Angell, suggest that corvids share seven unique capacities with humans: language, delinquency, frolic, passion, wrath, risk-taking, and awareness.

Not everyone agrees with Marzluff and Angell, but certainly no one disputes that smart birds and smart primates both possess curiosity. "When I was a teacher," Laura Erickson recalls,

"I always had a bell on my desk. It was from a board game, and it had a button on top. At the end of the school year, I would bring this bright-orange bell home and ring it in front of a blue jay I was raising. So he looks at the bell, his crest goes up, he jumps, and he flicks his wings and cocks his head. I ring it again, and he immediately goes up and pushes the button with his bill. But it doesn't make the cool 'Ding!' because his breast was pressed against the bell. He couldn't make it ding. He looks at me, and I ring it again. I leave and go to another room. All of a sudden, I hear this ding! Ding, ding, ding, ding! And so I come running, and he is hovering over the bell so he can press the button without his body touching it. Now what, in evolutionary terms, could possiby be of value in ringing a bell?"

The short answer: There is none. But as Erickson points out, "curiosity, tenaciousness, figuring out cause and effect, and keen observational skills are excellent qualities for an opportunistic omnivore, be it a blue jay or a human being." It is that kind of intelligence, she adds, "that gets my husband doing the *New York Times* crossword puzzle every morning."

BIRDS SHOW THEIR SMARTS IN OTHER ways, too. For three decades, animal psychologist Irene Pepperberg, PhD, worked with an African grey parrot she named Alex, teaching him to identify and categorize objects based on color, type, and texture. Alex was able to compare and contrast these characteristics, which, in Emery's view, provides "the best evidence that birds can form abstract concepts." While pigeons lack the linguistic ability of parrots, researchers have trained them to discriminate among visual images, to distinguish, for example, between photos that do and do not include images of humans, and to categorize paintings by Picasso, Monet, Chagall, and Van Gogh.

The key, researchers say, is to look beyond our biased view that whatever birds do, they do robotically. Navigation is a perfect example. We marvel at the supreme feats of avian navigation—at the Arctic tern, for example, which logs 50,000 miles on the wing each year, the equivalent, over a lifetime, "of three trips to the moon and back," says Jennifer Ackerman, author of *The Genius of Birds*. This ability was once thought to be innate, simple instinct. But as Ackerman

DURING MIGRATION, SNOW geese travel in large flocks and stick to fairly narrow routes that provide winds to follow, good visibility, and precipitation-free periods.

explains, "Now we know that bird navigation involves sensing, learning, and, above all, a remarkable ability to build a map in the mind."

And that goes for everyday navigation as well. A great many birds hide and then retrieve seeds they will depend on when food becomes scarce. Clark's nutcrackers, for example, can hide up to 33,000 pine seeds each autumn. Even hummingbirds must remember which flowers they visit and when, because after they extract nectar, it takes time for the flower to produce more. Western scrub-jays also combine navigation with memory. They remember not only where they stashed their future meals but also what they stashed there and when. This is important because the bird collects "not only nuts and seeds but fruit, insects, and worms—foods that perish at different rates," Ackerman explains.

And as if that weren't enough, some birds also face another challenge: Fellow birds that raid their hidden stores. As Emery explains, "caching behavior may be viewed as an arms race between storers and pilferers in which the storers use counterstrategies to minimize the risk of having their [food] pilfered." Storers might wait until other birds are distracted or have their view obstructed. They might make "false" food caches that contain either a pebble or nothing at all to mislead potential pilferers, or return alone to re-store food that was originally stored in the presence of others. It gets even more complicated, Emery says, because in this arms race, individual birds can play both roles—storer and pilferer. In fact, it pays to play both. Only birds that are also thieves employ anti-theft strategies. It takes a thief to know one.

STORERS MIGHT MAKE "FALSE" FOOD CACHES THAT CONTAIN EITHER A PEBBLE OR NOTHING AT ALL TO MISLEAD POTENTIAL PILFERERS.

WE ASSOCIATE BIRDS WITH SINGING almost as closely as with flying. And learning plays an important role here, too. "All birds produce calls, but not all birds sing," says Emery. "Young songbirds learn their songs from a tutor, usually the bird's father"—in many cases, only males of the species sing. "Perhaps surprisingly, humans learn language via a similar process to birds, during a sensitive period outside of which language doesn't develop properly." Ackerman writes, "If we can grasp what's going on in a bird's brain as it masters a melody, we might get a better handle on how our own brains learn language, why it gets harder over time, and maybe even how speech evolved in the first place."

Birds may even have something to teach us about music appreciation, according to Ofer Tchernichovski, PhD, a professor at Hunter College in New York City. "Birdsong is not only about attracting mates and defending territories. In some highly social songbirds, songs can be 'friendly.'" Given the beauty of birdsong, Tchernichovski asks, "Can songs elicit emotions in other birds, as music does in humans? We don't know yet, but, like music, birdsong is a cultural phenomenon."

The flexibility of birdsong—how it varies geographically—is also an object of study, as is mimicry. Birds

KNOWN FOR THEIR SMARTS, blue jays can mimic the calls of hawks to let other jays know a hawk is nearby. Opposite: The bare-throated tiger heron hunts by waiting motionless for fish, frogs, or crabs to come within reach of its long bill. This one is using a seed to lure prey.

EURASIAN HOOPOES ARE adept at using their long, curved beak to probe dung heaps and piles of dead leaves for beetles and other insects to eat. Opposite, top: Barn owls have a complex communication system. Opposite, bottom: A red- billed blue magpie can use its wedge-shaped beak to open shells.

renowned as mimics include mockingbirds, starlings, mynah birds, magpies, and lyrebirds. And the range of what's imitated is enormous: the calls and songs of other species, but also barking dogs, croaking frogs, buzzing insects, human voices, and even sounds including drinking water and trucks backing up.

Birdsong can also serve as a travelogue, says Erickson. "When you hear a mockingbird, you can tell where he's been by the songs in his repertoire," she says. "I call them the Othello bird because Othello attracted Desdemona by telling her tales of his adventures, and that's sort of what a mockingbird or brown thrasher or catbird is doing when they throw in these imitations: 'Look, I encountered a chainsaw! Listen, I once faced off with a cell phone!'"

CAN BIRDSONG ELICIT EMOTIONS IN OTHER BIRDS, AS MUSIC DOES IN HUMANS?

Scientists warn against anthropomorphizing the behavior of animals by interpreting their every action and reaction in human terms. But there is also danger in the opposite view, which de Waal calls "anthropodenial: a blindness to the human-like characteristics of other animals, or the animal-like characteristics of ourselves." It represents the age-old impulse to separate ourselves from the rest of the animal kingdom. The challenge we face now, as we learn more about the cognitive capacity of animals other than ourselves, says de Waal, is to ask ourselves: What kind of risk we are willing to take—the risk of underestimating animal mental life, or the risk of overestimating it? ●

Jurassic Origins

Birds are evolutionary descendants of dinosaurs, which helps explain why they are among the most diverse creatures on earth.

BY KAREN ASP

STRUTHIOMIMUS DINOSAURS, rendered at left, resembled an ostrich or emu. In 2004, researchers found hair-like feathers pressed into the rock alongside the bones of an early tyrannosaur species, making birds' links to dinosaurs even clearer.

When was the last time you saw a dinosaur in real life? It sounds like a crazy question, given that dinosaurs are extinct. Or are they? "Dinosaurs are alive and thriving," suggests Jingmai O'Connor, PhD, associate curator of fossil reptiles at the Field Museum in Chicago, adding that they outnumber species of mammals more than two to one. "They live in extreme environments and are capable of incredible physical feats like flying over the Himalayas and migrating for thousands of miles, and they have advanced brains capable of learning song and dance that's otherwise limited to humans."

Lest you think that the only way you can see dinosaurs is by watching *Jurassic Park* or visiting a museum, look out your window. See any birds? While you might call yourself a bird watcher, it might be more accurate to say dinosaur watcher.

The theory that modern-day birds descended from dinosaurs dates to the 1800s, and recent fossil findings have essentially confirmed this

PALEONTOLOGISTS VIEW Archaeopteryx (left and above) as a transitional species between dinosaurs and modern birds. The feathered dino-bird could fly short distances and in bursts. It was almost two feet long—a little larger than a green heron.

fascinating link. In 1861, a primitive bird fossil was discovered in Germany. Although the fossil, which was named Archaeopteryx, had sharp claws on its hands and a long bony tail like a dinosaur, it also had feathers and wings. "It looked like a cross between a bird and a dinosaur, and it was the first sign there might be a special relationship between dinosaurs and birds," says Steve Brusatte, PhD, a professor of paleontology and evolution at the University of Edinburgh in Scotland and the author of *The Rise and Fall of the Dinosaurs.*

Several years later, Thomas Henry Huxley, a famous biologist and close friend of Charles Darwin, noted similarities between Archaeopteryx and small dinosaurs, especially a two-legged group known as theropods, which included Tyrannosaurus rex. In the 1870s, Huxley suggested that theropods would be a likely candidate for the evolutionary transition from reptile to birds. "This would mean that birds are dinosaurs, because to evolve from a group means you still belong to it," O'Connor says.

Huxley's theory, though, never took hold, and although paleontologists resurrected the idea in the 1960s, based in part on a new study of Archaeopteryx, ornithologists rejected it. Then, in the mid '90s, a fossil find in China changed everything. Thousands of dinosaurs with feathers were discovered, convincing the skeptics. Today, as Brusatte says, "the theory that birds originated from dinosaurs is widely accepted."

If it feels like a stretch to think of birds as dinosaurs, consider the parallel of bats being categorized as mammals. Bats have hair like other mammals, give birth to live young, and possess other essential chracteristics of mammals. "Bats are part of the mammal family tree, and birds are the dinosaur equivalent of that," Brusatte says. "They're a strange type of dinosaur that got small, evolved wings, and, over millions of years of evolution, developed the ability to fly."

THE FIRST BIRDS APPEARED DURING a geologic period known as Jurassic, anywhere from 165 to 155 million years ago. Yet while they might have been identifiable as birds of some sort, they would have looked significantly different than the birds at backyard feeders today. "They were birdlike for dinosaurs, but not birdlike compared to living birds," O'Connor says. That's because they had teeth, claws, and long reptilian tails.

Birdlike dinosaurs had wings to power themselves through the air, and the feet of a T.rex or velociraptor resembled those of a chicken, O'Connor says. Birds and closely related dinosaurs also share soft tissues such as complex, vaned feathers arranged like a wing on the forelimb and, in some species, the hind limbs.

Thanks to fossils that have been unearthed throughout the decades, researchers have identified common behaviors, O'Connor says. Take, for instance, Mei long, a tiny troodontid dinosaur that's preserved sleeping with its head tucked under its wing like a bird, or Citipati, another type of dinosaur that was preserved while incubating its eggs.

These birdlike dinosaurs even had color, another finding researchers have gleaned from fossil discoveries. In well-preserved fossils or skin, researchers can see tiny bubbles, or melanosomes, that hold melanin or pigment. Although melanin doesn't create flamboyant shades like blue, red, and green, it is responsible for browns, whites, blacks, and ginger, which these birdlike dinosaurs had. While some feathers were shiny, others were iridescent, and still others were patterned or even ringed. "These early birds would have been a colorful bunch of characters," Brusatte says.

WHEN AN ASTEROID HIT Earth 66 million years ago, only feathered dinosaurs that had downsized to about 1 kilogram—the birds—were able to survive. Researchers theorize that their smaller size allowed them to adapt more easily to rapidly changing conditions.

ABOUT 100 MILLION YEARS AFTER birds made their debut at the end of the Cretaceous period, they had evolved from their dinosaur ancestors. Birds were a diverse, thriving species, and although primitive birds, those with teeth and claws, still existed, there were newer birds, too, including those with beaks. "A beak has evolved many times in birds, with the earliest known beaked bird living around 130 million years ago," O'Connor says.

Beaks weren't the only evolutionary development. Although dinosaurs were warm-blooded, birds were becoming more hot-blooded, a change that could have started to occur nearly 300 million years ago in various species, and perhaps a reason birds have been so successful, allowing them to thrive in extreme environments and adapt to change. "This means that the period of vulnerability early in their life is reduced, increasing their chances of survival," O'Connor says.

The Cretaceous also marked two other important transformations

in birds, namely their skeleton and brain. Today's birds have hollow skeletons that are pneumatic, meaning that they contain air sacs. "While dinosaurs had some skeletal pneumaticity, it's extreme in birds, helping them adapt to flight by reducing the weight of their skeleton," O'Connor explains.

And although birds' dinosaurian relatives had larger brains than other dinosaurs, birds' brains were becoming even larger. Some evidence suggests that birds had modern or modern-like brains after the first 50 million years of their evolution, but this hypothesis is based on a fragmentary picture of what birds were like during this period. Fossil records exist for the Archaeopteryx from 150 million years ago and several birds 130 to 120 million years ago, but then there's no record until 100 million years ago. All in all, though, it was a good time for birds—until, that is, catastrophe struck, changing the evolutionary devlopment of every animal, birds included.

BIRDS SEPARATE THEMSELVES INTO DISTINCT SPECIES BY EVOLVING DIFFERENCES IN COLOR AND SONG.

Around 66 million years ago, an asteroid believed to be six to nine miles wide smashed into Earth with a force of more than a billion nuclear bombs, punching a hole in the planet's surface that was more than 100 miles wide and triggering wildfires, tsunamis, and earthquakes at epic levels. The force not only obliterated the dinosaurs but also most of the primitive birds, those with teeth and claws. "It doesn't mean these birds had been unsuccessful or were failures—they were doing well," Brusatte says. "Yet when the asteroid hit, the rules of the game changed, and these types of birds that were once thriving died."

What remained was a largely empty world. Estimates suggest that 75 percent of the animals died. "After this mass extinction, there was opportunity and open space, allowing birds to evolve and diversify even more," says Brusatte, adding that many of today's major groups of birds, such as songbirds and raptorial birds, came out of this period.

But why did these birds survive the asteroid when so many others perished? There's no definitive answer, although researchers believe it was most likely a confluence of variables, which Brusatte explains with a gambling metaphor. "When the asteroid hit, it turned Earth into a giant casino with

SOME BIRD SPECIES HAVE retained obvious physical connections to their dino ancestors, such as the aptly named shoebill (opposite), with its foot-long bill. The cassowary (seen here) is thought to be more similar to ancient dinosaurs than most other birds. With its large frame and sharp claws, this flightless bird also has a casque, a helmet-like structure on its head, which many dinosaurs are believed to have had.

all of these animals playing poker," he says. "You're not going to win poker by having one card, but rather, a good hand of cards, which is how these birds survived." The "cards" that aided their survival included being small, having the ability to grow and reproduce quickly, and having a beak so they could eat seeds—which was critical. "When the animals and trees died and ecosystems collapsed, seeds would have been one of the few sources of food available," Brusatte says.

TODAY, THERE ARE MORE THAN 10,000 species of birds. While it's nearly impossible to know how many existed in prehistoric times, Brusatte says this is probably close to the maximum number of unique species that have ever lived at one time.

Several reasons explain the diversity. For starters, birds are able to separate themselves into distinct species simply by evolving differences in color and song. Their complex mating rituals may also play a role. "These rituals help promote speciation driven by sexual selection, which is obviously a major driver in the evolution of birds," O'Connor says.

When Darwin devised his theory of natural selection, he was perturbed by species that didn't fit his idea of survival of the fittest—he wrote in a letter to a friend that the tail of the peacock made him sick, for instance—which led him to propose sexual selection as another driver in the evolution of birds. "Looking at the enormous

ACCORDING TO RECORDS AND research, as many as 14 kinds of dinosaurs may have evolved into the modern-day ostrich (below). Seen in the evergreen forests of South Asia, the helmeted hornbill (right) looks quite like a small dinosaur. With a bright-red casque and solid beak, the bird is rare and endangered. Poachers and traffickers target it for its casque, which is softer than ivory and easy to carve into beads and figurines.

diversity of just tail morphologies in birds, with so many obviously not adapted for flight, you can readily see the powerful influence of sexual selection on bird evolution," O'Connor says. Plus, with their hot-blooded physiology, birds can live in almost any environment, and their flexible digestive system allows them to feed on an enormous variety of resources.

Besides satisfying our curiosity, what purpose does knowing the details of birds' origin serve? "I hope that this understanding provides a greater appreciation for birds and nature in general, which is desperately needed in this time of environmental crisis," O'Connor says. "Without an interest in the past, we cannot look to the future, and right now the future is bleak for our planet."

While there are more *types* of birds now than at other points in history, the total number of birds is on the decline. With the exception of wetlands, bird populations are trending down in every habitat, according to the 2022 U.S. State of the Birds report by the Audubon Society. From 1970 to 2019, 90 species lost at least 50 percent of their population.

If you would like to help scientists get a more accurate feel for how birds are faring, you can share sightings on eBird.org and participate in the Great Backyard Bird Count (birdcount.org), which takes place every February. No matter how you participate, remember this the next time you see birds in your backyard: You're observing the descendants of dinosaurs, who have existed for millions of years and, with the right circumstances, could have just as many more ahead of them. ●

DARWIN IDENTIFIED 13 FINCH species in the Galápagos Islands that were primarily differentiated by beak size. The finches share a common ancestor that arrived on the islands a few million years ago.

DARWIN'S DISCOVERY

What the great scientist found in finches led to the theory of evolution.

WHEN CHARLES DARWIN AND the crew of the *HMS Beagle* set foot on the Galápagos Islands in 1835, the English naturalist discovered a group of birds that would shape his groundbreaking theory of natural selection. Darwin found different finch species on the islands that were closely related to one another, but showed wide variations in beak, body size, and feeding behavior. He theorized that the differences in the size and shape of the beaks had enabled the different species to specialize in a range of foods, such as seeds, insects, cactus flowers and fruits, or even bird blood.

Darwin's notes and collected organisms from the Galápagos during his five-year voyage on the *Beagle* resulted in his theory of evolution by natural selection, one of the best-substantiated theories in the history of science. He collected several finch species, including the sharp-beaked finch, warbler finch, ground finch, small tree finch, large tree finch, common cactus finch, and large ground finch.

The variations within the finch family led Darwin to believe that the finches had adapted to the islands and unique environments where they lived, which helped him develop the theory that species are not fixed. Instead, they evolve over time, passing on traits that would increase their chances of survival in generations to come.

"It is interesting to contemplate an entangled bank, clothed with many plants of many kinds, with birds singing on the bushes, with various insects flitting about, and with worms crawling through the damp earth, and to reflect that these elaborately constructed forms, so different from each other, and dependent on each other in so complex a manner, have all been produced by laws acting around us," Darwin wrote in *On the Origin of Species* in 1859.

Darwin believed that evolution took place over hundreds or thousands of years and was impossible to witness in a single human lifetime. But Peter and Rosemary Grant, married evolutionary biologists who are currently emeriti faculty at Princeton University, saw evolution happen over the course of just two years.

Beginning in 1973, the Grants studied Galápagos finch populations for 40 years, examining evidence of natural selection in action. They found changes from one generation to the next in the beak shapes of the medium ground finches on the island of Daphne Major. The small island served as an ideal site for the research because the finches there had few predators or competitors, meaning that the main factors for survival for the medium ground finch are the weather and the availability of food.

Starting in 1977, Daphne Major received no rain for 551 days. Plants withered, and the finches grew hungry. The tiny seeds the medium ground finches were accustomed to eating were hard to come by. Finches with larger beaks could take advantage of alternate food sources because they could crack open larger seeds. The smaller-beaked birds died of starvation.

The Grants returned to Daphne Major the following year to document the effect of the drought on the next generation of medium ground finches. They measured the offspring and compared their beak size to that of the previous (pre-drought) generations. They found the beaks of the offspring to be 3 to 4 percent larger than those of their parents.

They had documented natural selection in action. "When we made the comparison between the size of the offspring generation and the population before selection, we found a measured, evolutionary response had taken place, and it was almost identical to what we had predicted," Peter Grant said at the time. Subsequent research by the Grants demonstrated the evolution of beak size in the species in response to other changing conditions on the island. The evolution occurred both in larger beaks, as in this case, and also in smaller ones when it was large seeds that became scarce.

Even though Darwin cracked the code on animal evolution nearly two centuries ago, his finches are still being studied. In 2018, evolutionary biologist Kevin Burns, PhD, and his postdoctoral student Erik Funk at San Diego State University published research that explored where the finches originated.

The study, published in *The Auk: Ornithological Advances*, presented a model to determine the ancestral biogeography (how a species' distribution varies over time and location) of Coerebinae, a subfamily of birds called tanagers, a group that includes what are now known as Darwin's finches and their 14 closest relatives.

"The results...were a bit surprising, because they suggested a dispersal pattern that was not necessarily the most 'straightforward' explanation for how these birds arrived in the Galápagos," Funk said at the time of publication. "I think one of the big takeaway messages here is the possibility that biogeographic events, like dispersal, may not necessarily happen like logic tells us they should. Darwin's finches are such a highly studied group, and it is often taken for granted they arrived from mainland South America, but hopefully our results show readers that there is no more support for this hypothesis than there is for a Caribbean origin."

1. Geospiza magnirostris.
2. Geospiza fortis.
3. Geospiza parvula.
4. Certhidea olivasea.

Birds of Prey

Also known as raptors, these hypercarnivorous birds hunt actively with their strong talons, sharp beaks, and killer instincts.

GOLDEN EAGLE

Widespread in the wilder areas of North America, Europe, and Asia, the magnificent golden eagle is a true predator. It searches for prey by soaring high or by scanning from its aerie. Once it has a meal in its line of sight, it plunges to capture it in its talons. A pair of golden eagles sometimes hunt together, with the second bird nabbing prey that evades the first. Although the golden eagle typically hunts small mammals like jackrabbits, some have been seen attacking large mammals and even fighting off coyotes or bears to defend their young or guard their prey.

BALD EAGLE

With its recognizable white head and dark plumage, the bald eagle has been the national emblem of the United States since 1782 and a spiritual symbol for Indigenous people for far longer. Bald eagles tend to scavenge more than they hunt and are known to go after other creatures' catches. If an osprey has captured a salmon, for example, a bald eagle may harass the river hawk until it drops the fish or snatch the prey directly from the osprey's talons.

SNOWY OWL

A powerful predator of the high Arctic tundra, the snowy owl is colored to be camouflaged amid the wintry terrain. Snowy owls will swallow smaller prey whole and are known to hunt a variety of food, including lemmings, Arctic hares, mice, ducks, and seabirds. Their powerful wings stretch four to five feet long and allow them to silently glide up behind prey and then rapidly accelerate.

VULTURE

Considered to be a part of nature's cleanup crew, vultures are the ultimate scavengers. They scour the landscape for deteriorating carcasses, helping to curb the spread of dangerous diseases and bacteria. Strong enzymes in their stomach protect them by killing off dangerous toxins and microorganisms. Vultures sometimes regurgitate on their would-be predators as a powerful defense mechanism.

PEREGRINE FALCON

The adult peregrine falcon uses its sharp beak to tear small pieces of meat and gently feeds them to its chicks. Peregrines often hunt other birds, such as pigeons in urban areas or shorebirds and ducks in more rural locations. Five to six weeks after hatching, the young are fully feathered and able to fly. They stick to their parents' territory for a few months and then are ready to survive and hunt on their own.

Self-Reflection

Do birds really know themselves? A cowbird offers insight. A renowned birder weighs in.

BY NOAH STRYCKER

I'll never forget, early one spring morning, being jolted awake by someone—or something—insistently tapping on the window.

TAP-TAP-TAP-TAP-TAP.

Who could it be? Was someone in trouble? I rolled out of bed, pulled on socks, and padded into the kitchen, where a sliding-glass door overlooks my back porch and yard. Sunlight streamed through the misty trees to a cacophonous accompaniment of dawn birdsong—but not a single person was in sight.

TAP-TAP-TAP.

The noise was startlingly loud and definitely coming from the glass door panel. As I groggily began to consider the possibility of a ghost—it must be a rather polite one to wait for daylight and knock at the door, I thought—I glanced down and noticed a small creature sitting on the porch.

It was a few inches tall, feathered black and brown, and fiercely alive: an adult male brown-headed cowbird, staring up at me with beady eyes. As I watched from inside, he stood on tiptoe, fluffed out his neck like a miniature powerlifter, and spread his wings in an aggressive display, then charged toward the glass and fluttered against it for a few seconds, dishing out a series of hard, echoing pecks. He had evidently taken his own reflection for another male cowbird trespassing on his territory, and he was determined to chase off the perceived rival.

ONLY TWO SPECIES OF BIRDS, pigeons and magpies, are considered to have passed the "mirror test"—the ability to recognize the bird in the mirror as themselves. This white wagtail, attacking its reflection in a car mirror, evidently perceives the image as a rival trespassing on its territory.

I put a screen outside the door to block the reflection on the glass, but within minutes the cowbird appeared at another window, clinging to the sill with his feet and stubbornly pecking his reflection's eyes out. Pretty soon he'd scouted every inch of reflective glass on the house, tilting at his own image like a feathered Don Quixote.

You'd think he would have eventually worn himself out, but Mr. Cowbird kept this up for hours. He returned the next morning, and, finding his two-dimensional opponents uncowed, went after them with renewed zeal. The house reverberated with the sound of birdbrained battles. I was besieged by a tiny bundle of attitude who was unable, apparently, to recognize himself in a mirror.

And as the cowbird's campaign stretched from days into weeks, it became clear that he would never get it. I couldn't very well block every window and turn the house into a cave, so I settled in to wait for his seasonal hormones to subside. Each morning after sunrise for the rest of the two-month mating season, he arrived as if on cue to defend his territory.

His dedication was as impressive as his thickheadedness was baffling. How could an intelligent animal like a songbird fail so utterly to grasp its own image? At least my cowbird wasn't alone. Every nesting season, all over the world, many birds fall into the same trap. In the U.S., cardinals, robins, and in some cases mockingbirds are known to attack reflections in windows. And all kinds of other bird species get fooled by mirrorlike surfaces. Once, I watched an adult male wild turkey wander close to a shiny chrome car bumper,

SOME RESEARCHERS BELIEVE the mirror test, which some magpies (above) have successfully completed, is outdated. Future tests might rely on mechanisms other than a mark to track self-awareness, since some species might be indifferent to a mark on their body and motivated by another measure. The brown-headed cowbird (left) distinguishes its kind from other bird species using a unique vocalization that only other cowbirds can produce and understand.

notice its reflection, and explode into full-on display mode: tail up, chest puffed out, wattles dangling. Even the side mirrors on cars can catch birds by surprise—sparrows and warblers sometimes perch on top of those mirrors in parking lots, lean over to peer at themselves, and mount a frenzy of territorial defense.

Given their unusual upbringing, cowbirds might be expected to have a better sense of who they are. The brown-headed cowbird is one of North America's most widespread avian brood parasites. Females lay their eggs in other birds' nests and leave their care entirely to hosts. This means that every wild cowbird is hatched and raised without its true parents, yet it somehow learns about its identity as a cowbird before mating season.

Cowbirds can learn, at least to some extent: Lab experiments show that cowbird chicks raised in isolation, deprived of interactions with adult birds, sing bizarre songs and even try to mate with the wrong species. But even if cowbirds get the mating part right in the wild, if they can't recognize their own reflection, can they really know who they are?

IN 1970, A YOUNG AMERICAN PSYchologist named Gordon Gallup Jr. introduced four wild chimpanzees to a mirror in a carefully controlled setting and discovered a whole new way to explore animals' sense of self-awareness.

To begin his now-classic experiment, Gallup kept each of his chimpanzees alone in a room for two days to get acquainted with their surroundings. Then he added a full-length mirror. Each chimpanzee was exposed to the mirror for 10 days while Gallup observed its behavior.

Gallup noted that all four chimps—two male and two female—responded to the mirror with great interest. Right away, they greeted their reflections with social displays, as if a stranger had dropped into the room. But after a couple of days, those behaviors changed.

BIRDS VIEW THE FIRST moving thing they see after hatching as their parent, a phenomenon called imprinting. At one aquarium, baby penguin chicks are fed and cared for by staff in penguin costumes to prevent the chicks from imprinting on humans. Here, an emperor penguin is with chicks in Antarctica.

The chimpanzees instead began to use the mirror for self-directed activities, grooming parts of their bodies that were hard to see. They even made funny faces at their reflections.

To evaluate whether the chimpanzees could recognize their own image, Gallup devised a clever test. He anesthetized each animal and painted an odorless dye on its forehead, then watched what happened when it woke up. Initially, not realizing they'd been marked, the chimpanzees paid no attention to their foreheads. But when they glimpsed their reflections in the mirror, the chimps' hands suddenly went up to wipe off the colored marks. The inference was clear: Because they'd used the mirror to remove undesired marks from their own bodies, the chimpanzees must have realized their reflections were, in fact, their own.

In the half-century since Gallup published his groundbreaking results in the journal *Science,* he and other scientists have tried the so-called "mirror test" with all kinds of creatures. A fascinating pattern emerged in primates. Besides chimpanzees, each of the other great apes—bonobos, orangutans, gorillas, and humans—passed the test. But despite repeated experiments, no monkeys—including baboons, marmosets, capuchins, tamarins, and macaques—ever got the hang of it. A fundamental divide seems to exist between apes and monkeys.

Virtually all other animals tested over the years have failed. In 2020, Gallup declared that none besides the great apes have "consistently and convincingly" demonstrated self-recognition in lab experiments with mirrors. But other scientists aren't so sure. Elephants, killer whales, and dolphins have shown particular promise, maybe unsurprising, given those animals' human-like qualities.

One compelling study documented a species of cleaner wrasse—a small tropical reef fish that specializes in removing parasites from other, larger fish—which tried to remove marks from its body by rubbing itself along hard surfaces after viewing its reflection. As far as claims of pet owners who insist that their dog, cat, or bird understands its mirror image, there is little scientific evidence to support that.

The only bird known to have recognized itself in a lab is the Eurasian magpie, a dapper black-and-white species with a long, iridescent tail. Together with the closely related black-billed magpie and yellow-billed magpie in the U.S., the Eurasian magpie is in the corvid family, along with crows, ravens, jays, jackdaws, rooks, nutcrackers, choughs, and treepies.

The corvids are often regarded as the smartest group of birds on earth,

MOST BIRD SPECIES IMPRINT ON THEIR PARENTS WITHIN DAYS OF HATCHING, FOLLOWING THE ADULTS AND COPYING THEIR BEHAVIOR.

at least according to our human ideas about intelligence. In 2008, three out of five magpies in a German experiment used a mirror to scratch at marks on their chins, passing the mirror test in a great leap for birdkind. Those results have not been replicated, but it only takes one bird to demonstrate capability. Even brainy chimpanzees only pass the mirror test 75 percent of the time.

The magpie result seems to fit a broader context of how people understand mirrors. Human babies don't figure out their reflection until about age 2, when they also start to have abstract emotions like sympathy, guilt, pride, and embarrassment—which require the realization of a self that is separate from others. Do birds, or corvids, at least, have similar feelings?

AS FOR COWBIRDS, NOBODY IS HAVING any philosophical discussions about their cognizance. At best, the brown-headed cowbird gets grudging respect for being successful. The bird has expanded far beyond its historic range in the Great Plains grasslands, where cowbirds probably evolved a nomadic lifestyle to follow wandering bison herds—with the need to dump eggs in other birds' nests.

Today, brown-headed cowbirds thrive from coast to coast, inhabiting all manner of fragmented forests and suburbs. They flourish in landscapes that have been cleared for agriculture, where they hang around livestock rather than bygone buffalo. They still sneak their eggs into the nests of unknowing foster parents. This seems like a terribly lazy and churlish thing to do, and people often judge them with contempt. But cowbirds offer a fascinating study in how birds develop a sense of who they are, because every cowbird chick experiences an identity crisis right out of the egg.

Most of the world's bird species imprint on their parents within days of hatching, following the adults and copying their behavior. The most extreme examples are in precocial birds, like ducks and geese, that open their eyes and leave the nest within hours. Ducklings imprint on the first large object they see, which is usually their mom. They can imprint on people, too, or even machines. I heard about a ruffed grouse chick that thought a farm tractor was its mother, followed it up and down the fields, and later tried to court it.

Altricial birds, including most songbirds, which hatch helpless and naked, also imprint on their parents. Before opening their eyes—even before leaving the egg, in some cases—most songbird chicks listen intently to the sounds around them. They memorize their parents' songs as a template for later developing their own vocal repertoire, and chemical changes in their brains crystallize this information for life.

Not cowbirds, though. A baby brown-headed cowbird's instincts are wired a little differently. As it grows, the chick begs for food from its host parents but otherwise pretty much tunes them out. When they are big enough to fly, young cowbirds start sneaking off to roost at night in open fields where other cowbirds live rather than sleeping with their host families. The cowbirds somehow know to seek out their own species rather than copy the adults at the nest where they were raised, who cluelessly continue to feed them for weeks.

Still, cowbirds must imprint on adult role models, even if their process is a bit delayed. Some evidence suggests that young brown-headed cowbirds visually examine their own bodies to figure out who they should associate with. In one study, scientists kept a group of cowbird chicks in a soundproofed lab so the birds would have no chance to imprint on adults. Half of the chicks had their feathers dyed an unnaturally bright color with Sharpie pens, while the others matured up with their typical grayish-brown juvenile plumage. As 2-month-old adolescents, the chicks were introduced to two adult female cowbirds—one natural-looking and one colored with a bright Sharpie. The normal chicks spent most of their time with the normal female, while the colorful chicks gravitated toward the colorful female. The chicks must have taken their social cues from their own personal appearance.

AS WITH MOST BIRDS, SOUND IS important for cowbirds. Chicks instinctively respond to the loud rattle call of adults, like a secret password that unlocks their ability to imprint on their own species. When they hear the call, young cowbirds fresh from the nest suddenly start paying attention—and their brains begin absorbing the information they need to solidify their identity.

Cowbirds have a longer window to figure this out than other songbirds

THE EUROPEAN ROBIN IS the unofficial national bird of the United Kingdom and a common sight in many yards and gardens. Known for their rich, harmonious songs, robins have multiple vocal cords and can create many notes simultaneously.

do, but it's not unlimited. After a few months, if they don't find other cowbirds, youngsters may imprint on the wrong role model. In one experiment, cowbird chicks raised with canaries eventually learned canary-like songs and courted canaries for the rest of their lives. (What the canaries thought about it remains a mystery.)

It's one thing to know your species, though, and quite another to realize that you are one individual in a universe of others. These are two distinct states of mind. Maybe that's a key difference between cowbirds and magpies, as hinted at by the mirror test. If a cowbird can learn which species it belongs to but can never learn to recognize its own reflection, then its learning process might tend to be more instinctive and less self-aware than, say, a magpie's.

The more we study cowbirds and lots of other birds that aren't given much credit for intelligence, the more independent-minded they seem. For instance, using tiny tracking tags, scientists have discovered that a female brown-headed cowbird often hangs around the nest where she laid her egg for days afterward, keeping an eye on how things go with its host parents. If the host tosses the egg out, the cowbird will sometimes destroy the nest in apparent retribution, like a Mafia boss enforcing a pecking order. Who knows what she's thinking? Is this cold-hearted revenge, or is she following a hardwired script? To paraphrase the American philosopher Thomas Nagel, no person will ever truly know what it's like for a bird to be a bird. Mirror experiments offer a glimpse into how animals see themselves, but tests like this are also designed by us, for us, to compare another species with ourselves.

It could be that birds are just hung up on understanding how mirrors work. Or maybe bird consciousness flows in ways that we can't begin to fathom. Only Mr. Cowbird knows for sure, and he's too busy tilting at windows.

TAP-TAP-TAP-TAP-TAP! ●

Birds and Us

From the lovebirds we live with to the finches at the feeder, these are birds that spend time with humans.

SOOTHING SONGBIRDS

Tits, chickadees, and titmice constitute Paridae, a family of birds known for their short, stout, pointed bills and rounded wings. Chickadees and tits (like those seen here), curious and active birds, are often bold enough to take sunflower seeds right out of a person's hand.

CUDDLY COCKATIELS

Known for the crest of feathers on top of their head, cockatiels are small members of the parrot family native to Australia. They're one of the most popular pet birds in America, thanks to their sociable and outgoing nature.

FEEDING FINCHES

Hanging thistle feeders attract finches, pine siskins, juncos, buntings and other birds year-round. Finches like to eat nyjer thistle seed, a small, thin black seed from the African yellow daisy. With its high oil content, it is a nutritious source of energy for backyard birds and is one of the most popular types of birdseed.

THE WHITE PIGEON AT HOME

Pigeons were once used as town-to-town couriers, but now urban feral populations are considered a health hazard and a nuisance. Pigeons can be raised to be loyal and devoted, however, making them great pets for those who have the space. They need indoor habitats that allow for enough wing-flapping room. Outdoor rescued pigeons can be easy targets for hawks or cats if they take to the sky, so keeping them in a predator-proof aviary is safest. Pigeons are active pets but not overly demanding of their owners.

BONDING WITH BUDGIE

Often referred to by their full name, budgerigar, budgies are also called parakeets and are one of the most popular pets in the world. Affectionate and animated, these birds can mimic human speech. Often they can even learn to speak with greater clarity and broader vocabulary than larger parrot species, such as macaws and cockatoos.

NESTING KESTRELS

North America's most common falcon is also the smallest one. American kestrels are specifically attracted to habitats that also see human activity, such as pastures, parks, and urban areas. Their numbers have been declining in some areas in recent years. To combat that, conservationists and homeowners erect nesting boxes to encourage the birds to rear their young safely. These boxes are specially designed to provide kestrels with the required space and warmth to care for their offspring.

ANCIENT EAGLE HUNTERS

In the vast landscape of Mongolia's Altai Mountains, ancient Kazakh hunters on horseback used eagles to track their prey. The tradition was passed down through generations. Today, the practice has become a source of tourism revenue from visitors who pay to see the famed birds in action. Golden Eagle Festival is one of the biggest eagle events in Mongolia and includes hundreds of trainers. It has been held more than 20 years, a well-known event for thousands of tourists.

Watching and Learning

Bird watching teaches us patience and keeps us calm. The practice also supports the conservation of the rarest species.

BY MELISSA MAYNTZ

IN 2016, THE U.S. FISH AND Wildlife Service found that 28 percent of wildlife watchers took trips away from home to feed, observe, or photograph animals. Birders' average age is 49, with nearly half of them under the age of 45.

Bird watching may, at first glance, seem like a simple hobby. But what's simple about an early-morning excursion to a chilly mountain canyon in January, standing hip-deep in snow on the bank of a frigid river, binoculars trained on a patch of rocks with foam surging around them, just for a glimpse of a slate-gray bird with pale eyelids vigorously diving into the icy water? A birder getting their first look at an American dipper will understand.

Bird watching can encompass a range of activities with far-reaching effects for individual bird watchers. The pastime also makes an impact on society overall, including local and international economies, conservation, health care, agriculture, technology, psychology, and other fields. The better we understand and appreciate birds, the better we can see how interwoven they are in so many different aspects of our lives, even for those who may not watch birds closely themselves.

THE POSITIVE EFFECTS OF SPENDING time in nature are well documented. Taking a simple walk to listen to and watch birds can be beneficial to mental and psychological health by reducing stress, anxiety, and tension, which can in turn reduce blood pressure, as evidenced by a 2017 study published in the journal *BioScience*. Similarly, Christopher W. Leahy, who holds the Gerard A. Bertrand Chair of Natural History and Field Ornithology (Emeritus) at the Massachusetts Audubon Society, has noted that a love of birds can be "a source of personal joy, solace, and relief from the all too prevalent stresses of modern life."

As we work to identify new species of mystery birds, puzzling out their markings and making comparisons with our favorite field guides and other resources, we also stretch our mental acuity, keeping brain function sharp and improving cognition and memory, which can help us retain new

information. "Unlike memory functions that tend to decrease with age, expert knowledge often continues to accumulate as we get older. This makes it an area of strength in older adults that we may be able to harness to mitigate age-related memory decline," says Erik Wing, PhD, a postdoctoral fellow at the Baycrest Rotman Research Institute in Toronto. Wing (yes, that's his real name) is the lead author of a study using bird watching in correlation with overall memorization and mental organization.

The health benefits of watching birds extend beyond mental wellness, of course. Birding means physical exercise, as it encourages walking and hiking through neighborhoods, parks, preserves, and trails. The more intense the birding—chasing a rarely seen species or negotiating a challenging trail to find an elusive bird—the better the chance the excursion will build strength and endurance. Carrying gear such as binoculars, spotting scopes, field notebooks, and trail cameras adds to the physical demands.

BIRDING'S BENEFITS FOR LOCAL, regional, national, and international economies is clear. The money spent on birdseed, feeders, entrance fees to nature preserves, new field guides, binoculars, and other items amounts to billions of dollars annually, one small purchase at a time. According to the Outdoor Industry Association's latest report, an estimated $800 billion is spent each year on outdoor recreation in the United States, with birdwatching having an economic benefit of $41 billion. On an even larger scale, bird

watching–related expenses can add up tremendously, influencing employment rates and entire industries.

For example, birders frequently travel to festivals or distant destinations in search of new birds to observe. The hugely popular Rio Grande Valley Birding Festival, held in South Texas each November, offers visitors a chance to see many rare and special birds, such as the green jay, plain chachalaca, and buff-bellied hummingbird. Participants can immerse themselves in more than 120 field trips during the festival's five days.

Attending festivals means not only travel expenses such as flights, hotels, and rental cars but also paying for the expertise of local nature guides, eating in local restaurants, and picking up extra supplies from local stores. This means more jobs and extra taxes for local and regional economies, which helps support schools, hospitals, libraries, emergency services, and roads. The farther bird watchers travel, the more far-reaching those economic benefits can be.

Beyond the boost in tourism dollars, there are far more subtle ways that birds benefit our economies. Çağan H. Şekercioğlu, PhD, a biology professor at the University of Utah as well as Koç University in Istanbul, says he has observed more than 8,600 species. Through his studies of avian ecology, ecosystems, and ornithology, Şekercioğlu has seen just how economically undervalued birds can be. "People don't make the connection and think birds are just ornamental," he says.

A LOVE OF BIRDS CAN BE "A SOURCE OF PERSONAL JOY, SOLACE, AND RELIEF FROM THE ALL TOO PREVALENT STRESSES OF MODERN LIFE."

An example: Vultures reduce rabies-related deaths and the transmission of other diseases by consuming carcasses that harbor dangerous bacteria. This leads to lower health care and medical costs for impacted communities and improves employment with fewer days missed due to illness or deaths.

Şekercioğlu notes that insectivores

CONTINUED ON PAGE 80

AT THE SNETTISHAM RSPB Reserve in Norfolk, England (above), a large flock of knots roost during high tide. When spring tides coincide with migration patterns each year, the birds' numbers are at their greatest. Right: Comprehensive guides offer valuable insight into identifying different bird species.

1

2

3

4

1. The belted kingfisher, often first noticed by its rattling call, can be found throughout North America. With a preference for waterside habitats such as ponds, large rivers, marshes, and rocky coastlines, it scavenges for food by plunging into water and capturing small prey with its bill.

2. With a short tail and a long, spear-shaped bill, the eastern meadowlark can be heard in spring in the eastern U.S. and desert grasslands in the Southwest. The meadowlark nests on the ground and sings from exposed perches such as utility lines.

3. The only bird species that lives exclusively in Florida, the Florida scrub jay breeds in cooperative flocks: Every nesting territory has an adult pair and one to six additional caretakers—usually the pair's previous offspring.

4. Widespread east of the Great Plains, the Baltimore oriole is common in open woods in summer. Fond of sweet nectar and fruit, the bright-orange and black birds are easily lured to sugar-water feeders.

5. The northern cardinal can be found in woodlands, thickets, and suburban gardens throughout the eastern United States. Cardinals readily seek out bird feeders, where they favor sunflower seeds.

CONTINUED FROM PAGE 77

and carnivores, such as flycatchers and raptors, are valuable species in that they consume agricultural pests, which can help to ensure better crops and less need for expensive pesticides or rodenticides. Similarly, birds that pollinate, such as hummingbirds, can improve crop yields.

BIRDS AND BIRD WATCHING ARE INTIMATELY connected to overall conservation efforts. Different projects and initiatives to protect birds can also offer greater benefits to other wildlife and our own communities.

Wetlands are prime habitats for ducks, rails, shorebirds, raptors, wrens, sparrows, and many other species, for example, and wetland conservation projects provide clean, healthy habitats for birds. At the same time, other creatures—amphibians, reptiles, mammals, fish, and insects—also use the protected habitats.

When wetlands are protected, society benefits. Wetlands help filter water to protect groundwater supplies from toxic contamination and also serve as barriers to erosion and flooding, protecting nearby coastal communities from environmental catastrophes. Even small protected areas in urban and suburban communities provide green spaces for everyone to enjoy, regardless of whether or not a visitor is watching specifically for birds.

Other fields of study, including some of the most lucrative and vital sciences and technological innovations, benefit from birds and bird watching.

Psychologists, for example, study more than just how watching birds impacts our personal well-being. By studying the interactions of bird flocks and mating courtship rituals, psychologists can better anticipate group dynamics and apply new ideas to human interpersonal relationships and leadership strategies.

Climatologists look to bird migration timing, patterns, and seasonal range changes to pinpoint delicate areas of climate change. This can help formulate plans for more effective conservation of critical regions. Similarly, birds can provide environmentalists with data on water or soil contamination, as birds are often impacted by such contamination before other local changes are noticed. This could help speed up control and mitigation for oil spills or chemical leakage.

Isolated finch populations in the Galápagos sparked Charles Darwin's theories on evolution in the 1830s. Biologists today continue to study birds to better understand genetics and cell mutations, research that

THE J.N. "DING" DARLING National Wildlife Refuge on Florida's Sanibel Island is known for its migratory bird populations, such as the pelicans above. Left: Lapwings often build their nests in rough or broken ground to help camouflage the eggs. The birds lay clutches of four earth-colored eggs from March to early June.

may one day help develop cures for cancer or other diseases. Forensic science also uses birds in unique and productive studies. By studying vultures and other carrion-eating birds, forensic scientists can better understand decomposition and postmortem identification, potentially assisting law enforcement and criminology. Some studies have tracked the scavenging behaviors of vultures to find recently disposed bodies, which could potentially help with homicide investigations.

Computer engineers who have developed geotags and radio transmitters for ornithologists have been able to adapt that work for use in mobile devices, medical technology, and even spacecraft. Aeronautical engineers have studied birds in flight to better understand how their wings, feathers, and muscles operate, allowing for refinements in aircraft design for greater fuel efficiency, safety, and maneuverability.

In 2023, aerospace engineering researcher Samik Bhattacharya received a $441,000 grant to study bird wing morphing, which could improve the stability of unmanned aerial vehicles in strong gusts of wind.

With so many unique and widespread connections to different parts of our lives, culture, and society, bird watching is powerful and influential, an active part of our biodiversity, and a contributor to so many aspects of our lives. ●

Wacky and Wonderful

Quirky plumage, distinctive bills, and eccentric markings: These are not the kinds of birds you see every day.

SECRETARY BIRD

This expressive bird is distinguished by its long legs and the dramatic array of feathers on the back of its head. Its name is thought to derive from the 1800s, when male secretaries in Europe wore dark tailcoats and knee-length pants and carried goose-quill pens behind their ears. Native to Africa, secretary birds are found south of the Sahara Desert and stand about four feet tall.

VOGELKOP LOPHORINA

Known for its elaborate courtship display, the male Vogelkop lophorina fans out its feathers to form a deep-black crescent punctuated by bright-blue feathers—suggesting a smile and a pair of eyes—and hops around its female target. It is a close cousin of the greater superb bird-of-paradise, a species that looks very similar. The key difference is that male birds-of-paradise create an oval shape instead of a crescent during their mating ritual. The Vogelkop lophorina is native to Bird's Head Peninsula in New Guinea (*vogelkop* is "bird's head" in Dutch).

COCK-OF-THE-ROCK

Known as *tunki* in Quechua (an Indigenous language family spoken by the Quechua peoples, primarily in the Peruvian Andes), the Andean cock-of-the-rock is the national bird of Peru. Male birds have vibrant red-orange feathers in a massive fan-like crest that obscures a small bill. True to its name, the cock-of-the-rock prefers nesting sites near rocky cliffs and forest ravines.

KING VULTURE

Like other vultures, the king vulture glides on air currents, scavenging the corpses of dead animals from the land below. Unlike their vulture relatives, they have multicolored heads and necks. Unlike other colorful birds, king vultures are bald, which is believed to help prevent disease-laden animal remains from festering in dense plumage.

VICTORIA CROWNED PIGEON

Considered the peacock of the pigeon family, the Victoria crowned pigeon (left) is a large blue-gray bird with a deep maroon underbelly. Its crowning feature is its erect, fan-shaped crest of lacy feathers. Once hunted for this prized plumage, the birds are now protected by law and live in the jungles of New Guinea, far from human settlements.

PARADISE TANAGER

With a lime-green head that contrasts with its blue and black body (right), the paradise tanager hails from the Amazonian lowlands. While male birds of other species use their bright plumes in mating rites, both male and female tanagers boast multicolored feathers.

WILSON'S BIRD-OF-PARADISE

A fairly small bird native to the foothill forests on Waigeo and Batanta islands in Indonesia, Wilson's bird-of-paradise has quirky, curly tail feathers and a featherless turquoise head that resembles an exposed cranium. When British naturalist Edward Wilson brought the species to Napoleon's nephew, Charles Bonaparte, he named it after Wilson as a way to honor the Commonwealth.

FRIGATEBIRD

Beachgoers along the coasts of the southern United States, Mexico, and the Caribbean might notice this large black pterodactyl-like bird (left) soaring on the tropical breeze, using its forked tail to steer through the current. Male frigatebirds have a bright red pouch on their chest, which they inflate to attract females.

GOLDEN PHEASANT

With red and gold plumage that resembles the headpiece of an pharaoh, the golden pheasant (right) is a game bird native to the forests and mountainous areas of China. These birds are clumsy fliers, only taking to the air in short bursts. Instead they run and spend most of their time on the ground. They are regarded as a sign of luck and prosperity.

LIFE

Birds

Dotdash Meredith Premium Publishing

EDITOR IN CHIEF Kostya Kennedy
CREATIVE DIRECTOR Gary Stewart
EDITORIAL OPERATIONS DIRECTOR Jamie Roth Major
MANAGER, EDITORIAL OPERATIONS Gina Scauzillo
EDITOR Courtney Mifsud Intreglia
CONTRIBUTING WRITERS Karen Asp, David Bjerklie, Melissa Mayntz, Noah Strycker
PHOTO EDITOR Rachel Hatch
COPY CHIEF Tracy Guth Spangler
REPORTER Maya Kukes
PRODUCTION DESIGNER Sandra Jurevics
PREMEDIA TRAFFICKING SUPERVISOR Sarah Schuster
PREMEDIA IMAGING SPECIALIST Paige E. King
COLOR QUALITY ANALYST Sara Luckey

VICE PRESIDENT & GENERAL MANAGER Jeremy Biloon
VICE PRESIDENT, GROUP EDITORIAL DIRECTOR Stephen Orr
SENIOR DIRECTOR, BRAND MARKETING Jean Kennedy
ASSOCIATE DIRECTOR, BRAND MARKETING Katherine Barnet
SENIOR MANAGER, BRAND MARKETING Geoffrey Wohlgamuth
BRAND MANAGER, BRAND MARKETING Mia Rinaldi

SPECIAL THANKS Gabby Amello, Brad Beatson, Farah Ameen

Dotdash Meredith

PRESIDENT, LIFESTYLE Alysia Borsia

Dotdash meredith MPA THE ASSOCIATION OF MAGAZINE MEDIA

PRINTED IN THE USA

Photo Credits

Front covers: (cardinal) Rolf Nussbaumer/NPL/Minden Pictures; (oriole) phototrip/iStock/Getty; (background) Penpak Ngamsathain/Moment/Getty
Back cover: (from top) Péter Hegedus/500px Prime/Getty; Wahrmut Sobainsky/NiS/Minden Pictures
Endpapers: BOTOBOX/iStock/Getty

Page 1: Don White/E+/Getty **Pages 2–3:** B&S Draker/NPL/Minden Pictures **Page 4:** stanley45/iStock/Getty **Pages 6–7:** Darrell Gulin/The Image Bank/Getty **Pages 8–9:** Tetra Images/Getty **Pages 10–11:** Gerry Ellis/Minden Pictures **Pages 12–13:** Jonathan Ross/iStock/Getty **Pages 14–15:** Malcolm Schuyl/FLPA/Minden Pictures **Pages 16–17:** Freder/iStock/Getty **Pages 18–19:** Vicki Jauron/Moment/Getty **Pages 20–21:** Ernie Janes/NPL/Minden Pictures **Pages 22–23:** Dr. James St. Clair **Page 24:** (clockwise from left) Cyril Ruoso/NPL/Minden Pictures; Reiner Riedler/Anzenberger/Redux; Michael Quinton/Minden Pictures **Page 25:** Rick Friedman/Corbis/Getty **Page 26:** Neil Bowman/FLPA/Minden Pictures **Page 27:** (from top) Staffan Widstrand/NPL/Minden Pictures; Kim Taylor/NPL/Minden Pictures **Pages 28–29:** Spondylolithesis/iStock/Getty **Page 30:** Greg Basco/BIA/Minden Pictures **Page 31:** GummyBone/iStock/Getty **Page 32:** Thomas Hinsche/BIA/Minden Pictures **Page 33:** (from top) Paul Sawer/FLPA/Minden Pictures; eiffel/500px/Getty **Pages 34–35:** Mohamad Haghani/Stocktrek Images/Getty **Pages 36–37:** (from left) Dickson Lee/South China Morning Post/Getty; De Agostini/Getty **Pages 38–39:** Mark Garlick/Science Photo Library/Getty **Page 40:** Cyril Ruoso/Minden Pictures **Page 41:** Duncan Usher/Minden Pictures **Pages 42–43:** (from left) AndreaWillmore/iStock/Getty; Tim Laman/NPL/Minden Pictures **Page 44:** Tim Graham/Alamy **Page 45:** The Print Collector/Alamy **Pages 46–47:** Daniel Hernanz Ramos/Moment/Getty **Pages 48–49:** Alan Murphy/BIA/Minden Pictures **Pages 50–51:** skynavin/iStock/Getty **Pages 52–53:** JackF/iStock/Getty **Pages 54–55:** Luke Massey/NPL/Minden Pictures **Pages 56–57:** Wikipedia **Pages 58–59:** (from left) Irving A. Gaffney/iStock/Getty; phototrip/iStock/Getty **Pages 60–61:** Klein and Hubert/Minden Pictures **Pages 62–63:** Andrew_Howe/E+/Getty **Pages 64–65:** Anastasiia Krivenok/Moment/Getty **Page 66:** Ray Kachatorian/Stone/Getty **Page 67:** GarysFRP/E+/Getty **Pages 68–69:** Aleksandar Jankovic/E+/Getty **Page 70:** HMVart/E+/Getty **Page 71:** Gary Chalker/Moment/Getty **Pages 72–73:** Timothy Allen/Stone/Getty **Pages 74–75:** Pamela Moore/E+/Getty **Pages 76–77:** (from left) David Tipling/NPL/Minden Pictures; Luis Robayo/AFP/Getty **Page 78:** (clockwise from top left) Donald M. Jones/Minden Pictures; 1432VIJAY/Shutterstock; Brian Lasenby/iStock/Getty; All Canada Photos/Alamy **Page 79:** Jun Zhou/iStock/Getty **Pages 80–81:** (from left) Andrew Linscott/E+/Getty; Jeffrey Greenberg/Universal Images Group/Getty **Pages 82–83:** Mark Newman/The Image Bank/Getty **Page 84:** Tim Laman/NPL/Minden Pictures **Page 85:** Thomas Marent/Minden Pictures **Pages 86–87:** miroslav_1/iStock/Getty **Page 88:** Rod Williams/NPL/Minden Pictures **Page 89:** Edward Myles/FLPA/Minden Pictures **Pages 90–91:** Otto Plantema/Buiten-beeld/Minden Pictures **Page 92:** RainervonBrandis/iStock/Getty **Page 93:** fenlio/iStock/Getty **Pages 94–95:** Roland Seitre/Minden Pictures **Page 96:** Vicki Jauron/Moment/Getty

NICOBAR PIGEON

Native to Indonesia, Nicobar pigeons have iridescent feathers in shades of green, yellow, copper, and blue. In the shade, they look to be a muted gray, but sunlight picks up the intense pigments of their dazzling plumage. They also boast a wreath of long neck feathers, known as hackles, which further contribute to their flashy appearance.

Taking Flight

"Everyone likes birds. What wild creature is more accessible to our eyes and ears, as close to us and everyone in the world, as universal as a bird?"

—DAVID ATTENBOROUGH

Made in United States
Orlando, FL
17 September 2026

87840233R00059